Waxy Corner

By

Jenny Gammon

Northern Bee Books

Waxy Corner
© Jenny Gammon 2016

Published in the United Kingdom by
Northern Bee Books,
Scout Bottom Farm,
Mytholmroyd,
West Yorkshire HX7 5JS
Tel: 01422 882751
Fax: 01422 886157

www.northernbeebooks.co.uk

ISBN 978-1-908904-89-8

Design and Artwork
D&P Design and Print
Worcestershire

Waxy Corner

By

Jenny Gammon

Contents

Salvaging Wax 5

Thinking of Making candles 7

Making Candles with Silicone Moulds 9

Making Candles with Rubber Moulds 11

Care of Rubber Moulds 13

Making Rolled Candles 15

Having fun with Rolled Candle 17

Dipped Candles 19

About the Author

Jenny has been Beekeeping with husband Sid for nearly 30 years. Both of them have been involved with the Taunton and District Division of Beekeepers in Somerset.

She got interested in wax after a talk at the local division.

She has done talks to other divisions and held workshops in making of candles.

She was awarded the West Country Honey Farms rose bowl in 2011.

She was made President of the Taunton Beekeepers in 2013.

Salvage Wax from discarded Brood /Super Frames

You really need a solar extractor to do the job properly.

A wooden box with a stainless steel insert –
Go on the internet to see designs on a beekeeping site

e.g. Thornes / National Bee Supplies etc.

Have some foil cooking containers to catch the wax - not expensive.

Put your old wax into discarded tights / stockings before putting them into the solar extractor. These will save you having to steam clean the solar extractor or use a lot of elbow grease to clean it.

You may have to melt down the wax you have taken off to clean it again if it is for something special.

If you decide that you really want to do it indoors :

You need two different size saucepans –one to fit inside the other. Most people have some old ones lying around. Make sure you have water in the lower saucepan before putting it on the heat – enough water to come up the side of the saucepan dropping into it. A little water in the above saucepan preferably distilled. Break up the wax in the upper pan. Wait for the old wax to melt. When melted allow to cool. The wax will float to the surface and you can throw away the dross below.

To get really good wax for special things use cappings after you have extracted your honey. Wash the capping in distilled water to remove the honey –then melt down the wax as above in saucepans. Remember the longer you heat wax the darker it becomes.

Distilled water: you can buy bottles of this or try using the water from your dryer / humidifier.

Remember not to leave the wax cooking – hot wax can cause a fire if it over flows (i.e. Just like lighting a candle – it burns.) Don't have children / animals around you at the time / don't answer the phone and forget what you are doing –accidents can happen especially if you are making candles and the mould tips over when you balance it and the wax is all over the floor / and kitchen worktops.

Put down paper before you begin anyway.

When you want to clear your wax and keep it for later -try lint attached to an ice cream container and pour into that. Put your old lint into the solar extractor next year and get the surplus wax out of them.

Remember that wax is MONEY – don't throw it away. You can make candles /wax flowers/ exchange for equipment or new foundation when you visit your beekeeping supplier.

Thinking of Making Candles:
What you will need to start

1. You will need a small stove (see photo) or be brave enough to use your cooker and find you have to clean off all the wax that you have dropped on it. Ceramic and wax do not mix. Making candles is not a clean job.

2. Two saucepans - one that will fit inside the other. Don't use your best kitchen saucepans —old ones from a junk shop will do as long as you have scrubbed them clean and the handles are firm and not likely to come off when you have hot melted wax in them. (see photo)

3. A Pyrex or glass jug - easier to clean than a plastic one (see photo)

4. Surgical lint. —This you can get at some pharmacies' or go on the internet –

5. Hopefully you still have a few clothes pegs around the house. You will need these to attach the lint onto the jug. —Make sure that the lint is fluffy side up.

6. You then have to select the wick you are going to use. This must be wick suitable for use with beeswax not with Paraffin candles. Wick can be purchased from your local

beekeeping supplier -generally in 5 m lengths. If you really want to go into business wick comes in 50m lengths.

Wick is generally supplied in ¼", ½", ¾", 1", 1 ¼", 1 ½", 1 ¾", 2", 2 ½", 3", 3 ½" 4".

To decide which size wick you should use, get a measuring tape and measure the diameter of the candle mould that you are going to use. A straight candle is easy to determine. If you have a fancy candle mould e.g. Dog, Cat etc. you will have to take a mean average. Most of these types of candles will not be burned, so you do not need to be exact. If you do not use the correct wick on straight candles / any candles, you will find they will either have wax running down the side of the candle or will pool by the wick and put out the flame.

7. You will also need a wicking needle - a long needle with a large eye.

8. Cocktail sticks to hold the wick in place once you have poured the wax.

9. Glycerine (cake making section in a super market) and also washing up liquid are also a necessity for Rubber moulds.

10. You need to get yourself a candle mould. Get yourself something simple to start. There are Latex moulds and also Rubber moulds – Latex moulds can prove quite expensive. Rubber moulds a lot cheaper. (Go on the internet).

11. A **FLAT** iron is very handy to flatten the bottom of candles. Not steam.

Making Candles with Silicone Moulds

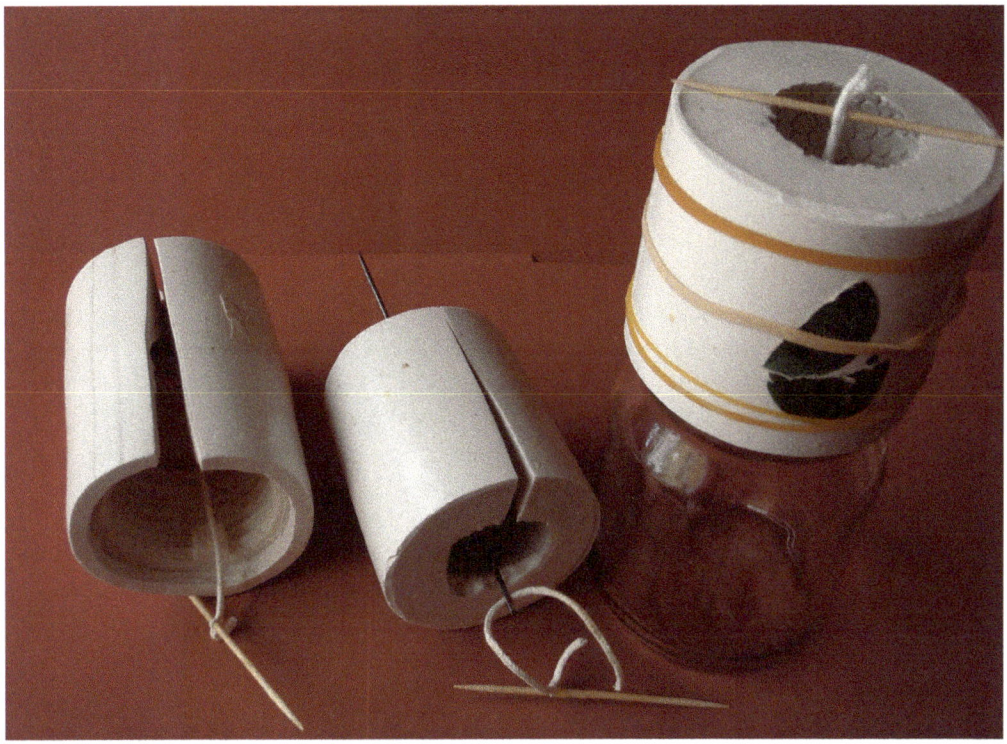

Silicone moulds are not cheap but handled with care can last a long time.

Some silicone moulds are split right through to the hole where the wick is placed. Other silicone moulds are split part way –do **NOT** split these up to the top.

To make these candles you will need rubber bands / cocktail sticks / wicking needle / wick/ and a bottle to rest the mould onto to let the wax set.

1. Using the moulds that are split all the way up > select your wick size – see wick sizes shown on previous pages – measure the diameter of the mould and use the wick needed. E.g. 1" diameter = 1" wick. 1 ½" diameter = 1 ½" wick etc.

    Cut the wick slightly longer than the length of the mould. Put the cocktail stick through the end of the wick > the cocktail stick goes across the open end of the mould. Run the wick through the mould and out through the hole at the top.

2. Using the moulds that are not split all the way up > Measure and select your wick size. Put the cocktail stick through the end of the wick. Thread a wicking needle (or a very long darning needle) and pass the needle through the hole at the top.

3. For both types of silicone moulds> Take your rubber bands and put around the mould so that the wax when poured cannot escape. Especially tighten the rubber band on the mould that is split all the way up.

4. It is a good idea before pouring your wax that you place the mould into your freezer for a couple of minutes. This gives the wax a lovely shine when you take the candle from the mould.

5. Place the mould onto an empty jar which should be standing on a flat surface and safely away from being jarred.

6. Pour your wax into the mould – (do not have the wax **boiling**) if the wax starts seeping out the other end into the bottle you have not tightened the rubber bands enough. Sometimes just squeezing the mould at that end allows the wax to harden.

7. When the wax has partially set – gently pull the wick through from the top end until the wick does not show on the bottom end of the candle. Allow the candle to set completely.

8. Remove the rubber bands and very gently remove the candle from the mould. With the moulds that are part way split –make sure the wax is completely set before trying to remove the candle. Otherwise you may find that you leave the wick behind.

9. Trim off the wick at the top of the candle to make it look good. If you have forgotten to pull / could not pull the wick through before it is set –you will need to trim off the wick on the bottom of the candle – then use an old type flat iron to smooth the wax.

10. Do not store your mould with the rubber band around them.

Making Candles with Rubber Moulds

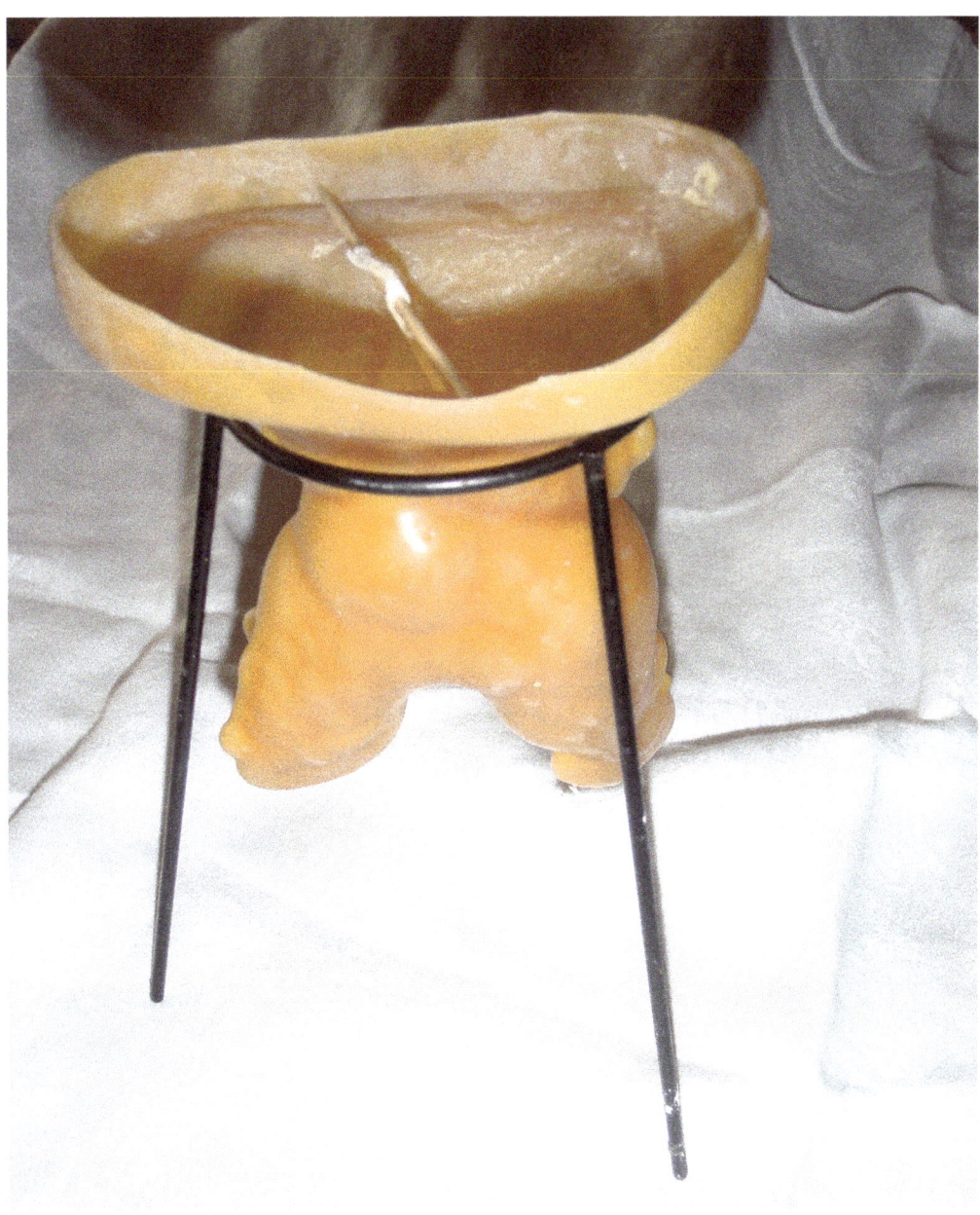

Rubber moulds are cheap and handled with care can last a long time. After using them —clean with warm soapy water, dry off and powder the inside. Stuff with kitchen paper and keep in dark place.

To make these candles you will need cocktail sticks / wicking needle / wick/ glycerine and a suitable stand to rest the mould onto to let the wax set. (example shown).

1. Turn the mould inside out and coat the inside with glycerine. Remove extra. Don't flood the mould. (Get glycerine from cake section of Super market).

2. Having selected the wick you require cut the wick slightly longer than the length of the mould. Put the cocktail stick through the end of the wick > the cocktail stick goes across the open end of the mould. Run the wick through the mould with a wicking needle and out through the hole at the top.

3. It is a good idea before pouring your wax that you place the mould into your freezer for a couple of minutes. This gives the wax a lovely shine when you take the candle from the mould.

4. Place the mould onto an empty jar / stand which should be standing on a flat surface and safely away from being jarred. (See picture of a stand)- If the candle can fall through the hole at the top use a piece of cardboard cut into a collar shape. Make sure you can get the candle out of the holder after the wax has set.

5. Pour your wax into the mould – (see previous to see clearing of wax – do not have the wax boiling).

6. Make sure that you have squeezed the mould slightly, especially if it has places where air can gather. Make sure that if there are any heads etc. you have turned the mould out correctly. You may end up with a headless "bird" candle.

7. When the wax has partially set – gently pull the wick through from the top end until the wick does not show on the bottom end of the candle. Allow the candle to set completely.

8. When you are ready to remove the candle from the mould –coat the outside of the mould with a little washing up liquid. This will allow the rubber to slide on itself. Remove the candle from the mould over a basin in case you have tried to remove the candle too early and it is not set.

9. Trim off the wick at the top of the candle to make it look good. If you have forgotten to pull / could not pull the wick through before it is set –you will need to trim off the wick on the bottom of the candle –use a old type flat iron to smooth the wax.

Care of Rubber Moulds

After using a rubber mould wash with warm water, dry carefully, leave for a short while to allow the inside to dry. Powder the inside with talcum powder. Make sure you do not pile one on top of each other and therefore loose the original shape.

If you find disaster has fallen and you have not done the above – hot water- not boiling – and wash the mould inside and out. When flexible dry carefully and immediately stuff with kitchen paper. You will find that the mould will regain shape –unless you have left it so long it has started to perish.

Leave for a while to set its shape –remove the stuffing and powder the inside. If necessary re-stuff the mould with a dry kitchen paper till you want to use it again.

Making Rolled Candles

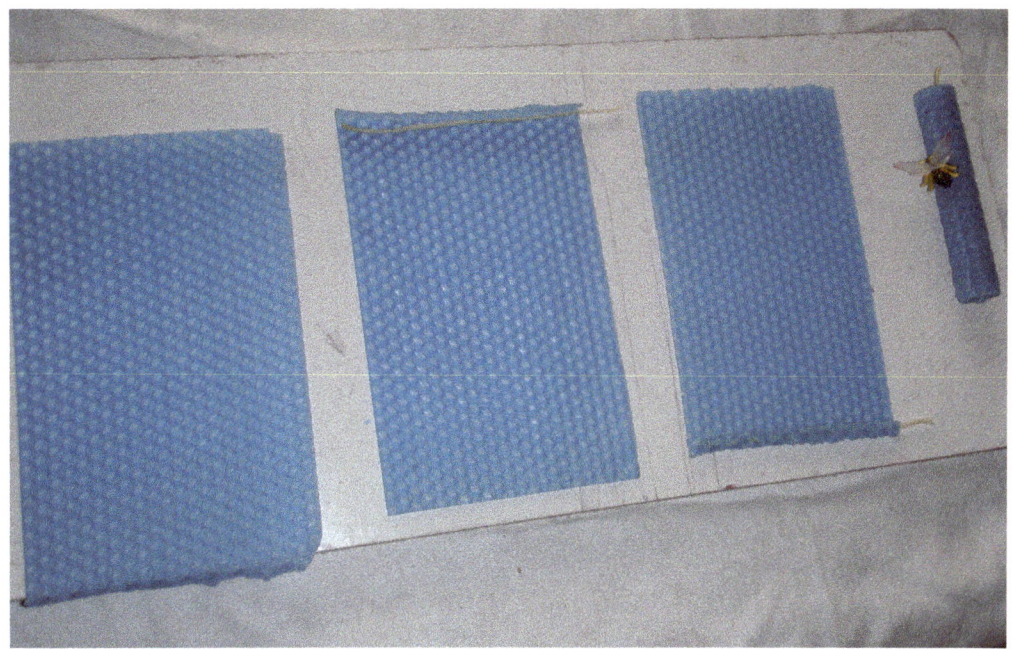

Please remember that foundation is brittle when cold. Store the foundation at room temperature before using or have a hair dryer to blow on the wax / bottle of warm water to roll over the wax.

You will need scissors to cut the wick / a Stanley knife to cut the foundation if you are dividing it. Hair dryer or a bottle of warm water. Also a board to roll the candle on.

If you are making candles to use at a show for children to "candle roll" I suggest that you cut the sheet of foundation in 4 pieces. Use a waxed length of ½ "wick.

If you are making full length rolled candles (1 complete sheet) you will need !" wick.

Remember that if you add extra foundation to make the candle thicker you will have to judge the width and use a thicker wick.

Making the candle:

1. Fold the edge of the foundation over the edge of the board. Figure 1
2. Turn over the foundation and lay the wick in the grove you have made. Figure 2

3. Make the foundation secure around the wick. Figure 3.

4. Start rolling the foundation to make your candle > watch one side so that you get the candle straight.

5. Press the end of the foundation to the rolled candle. You will see that the wax will adhere to itself. Finish off (maybe) with a plastic bee to improve presentation. Figure 4.

Having fun with Rolled Candles

Hat Candle

You will need 2 beeswax foundation sheets – choose your colour

Approx 3" of wick - I used a 2" wick

24" fused pearls

18" ribbon to match your foundation

3 wide peach silk roses or if you have got to that stage make your own wax flowers.

(Suggest trying to get "An introduction to Beeswax Flower Making" by Elizabeth Duffin)

Some dried baby's breath or plastic looking the same

Some greenery

Cut seven 2 ½ " strips and make a basic rolled candle – butt joining the edges.

Cut ten ½ " strips and wrap around the candle bottom forming the hat brim

I pinned the ribbon in place around the brim with a coloured headed pin allowing the tails to cross and extending on the right hand side. Trim the tails at an angle just along the edge. Wrap the pearls around the hat as you want.

Tuck the flowers /baby's breath and greenery into the ribbon or glue them in place.

Dipped Candles

Not the easiest candles to make.

You will need

A dipping bath -I have known this to be a couple of Baked bean tins > bottom of one cut out and both soldered together. Also something to put the dipping bath into that will heat the water to heat the wax in the dipping bath. The water should come at least half way up the dipping bath. Put a couple of piece of wood under the bath so it is surrounded by water.

You will need a lot of wax in your dipping bath and also you will need to have extra wax warming on the side to top up the bath. Every dip you do takes away some of the wax from the bath. The wax in the dipping bath will take an hour plus to melt –unless you have put in melted wax.

I have a dipping "Hold" that can do eight candles at once and as there is tension on the wick have found that I can even make the candles outside. I will explain later. You attach wick to the "Hold" to the width of the candle you wish to make - you can make up to 1" wide candles. When you get to this size and need wider candles - remover the candles and then dip them further by hand to make them wider. Remember that on the first dip you must leave the "Hold" in the wax to allow the air to escape from the wick. (See photo)

You have to use a rhythm when dipping. If you stop at the bottom to think about things the heat of the wax will remove what has already adhered to the wick. I use a "going down" one-two-three-four and then "start up" with a one–two–three-four all in one sequence. If you don't have a rhythm you will get wax running down the side and not get a smooth candle. Then hang the "Hold" out to dry for a few minutes - Doing the hovering in between is a way of giving wax time to dry. Too quick between dips and you are going to get a candle wide at the top and narrow at the bottom where it has melted away.

Once you have got to your candle thickness allow to dry and when almost cold remove from the hold.

You could try to make your own "Hold" for two candles. Use dowel and wire / old wire hanger. That would cut the cost to about £1. (See picture).

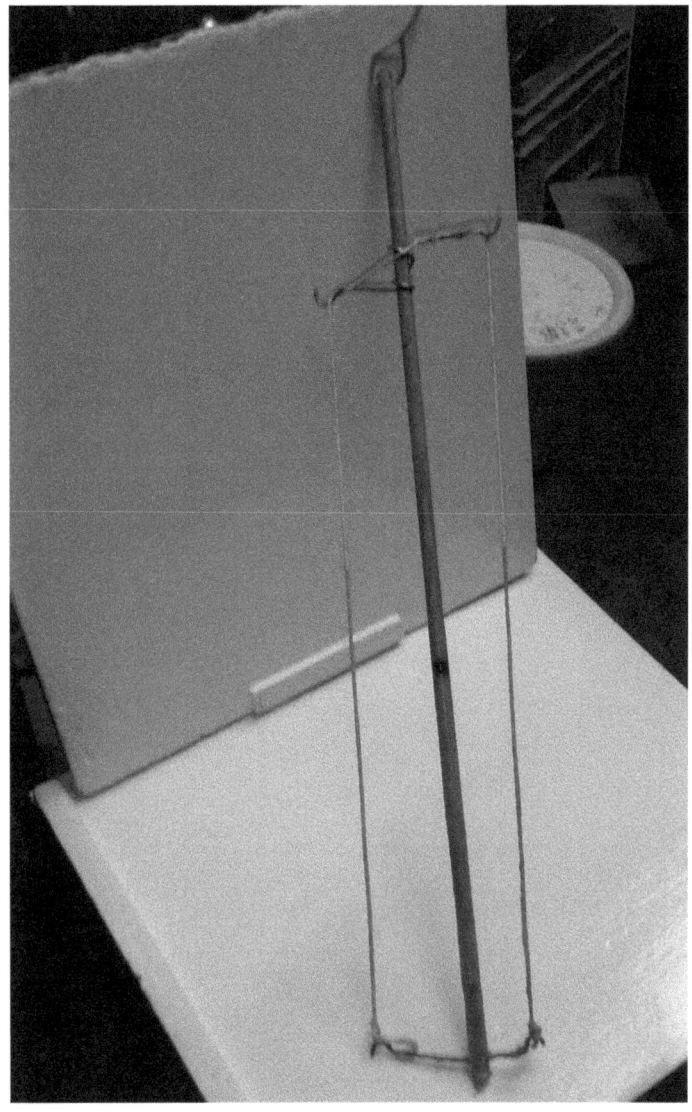

Second way

You can make dipped candles in pairs with just holding the wick – dipping and then hanging them up, using clothes pegs to secure. Again you have to allow the air to rise out of the wick as above. Remember to pull the candle straight after each dip till you have a fair bit of wax on the wick.

Then you have to leave the candles in between to dry off – after a few dips you need to judge if the wax is the right temperature –get two sheets of glass –place the candle between the two sheets of glass and roll them straight. You will probably have to do this a few times while making the candles.

Problems I have found:

When hanging the candles up to dry you need a place that is draft free. Someone usually makes this impossible by walking through. The candles then decide that they are going to bend slightly.

When you go to roll them between glass – I could never judge the right temperature and ended up with more wax on the glass than on the candles- With regret I gave up on that method.

www.ingramcontent.com/pod-product-compliance
Lightning Source LLC
Chambersburg PA
CBHW041632040426
42446CB00022B/3486